Fun on the Farm

AT THE SHEEP FARM

By Bruce Esseltine

Gareth Stevens
PUBLISHING

Please visit our website, www.garethstevens.com. For a free color catalog of all our high-quality books, call toll free 1-800-542-2595 or fax 1-877-542-2596.

Cataloging-in-Publication Data

Names: Esseltine, Bruce.
Title: At the sheep farm / Bruce Esseltine.
Description: New York : Gareth Stevens Publishing, 2017. | Series: Fun on the farm | Includes index.
Identifiers: ISBN 9781482455328 (pbk.) | ISBN 9781482455342 (library bound) | ISBN 9781482455335 (6 pack)
Subjects: LCSH: Sheep–Juvenile literature. | Sheep farming–Juvenile literature. | Sheep ranches–Juvenile literature.
Classification: LCC SF375.2 .E87 2017 | DDC 636.3–dc23

First Edition

Published in 2017 by
Gareth Stevens Publishing
111 East 14th Street, Suite 349
New York, NY 10003

Editor: Ryan Nagelhout
Designer: Laura Bowen

Photo credits: Cover, p. 1 brackish_nz/Shutterstock.com; p. 5 CatJB/Shutterstock.com; p. 7 branislavpudar/Shutterstock.com; p. 9 TOM KAROLA/Shutterstock.com; p. 11 1000 Words/Shutterstock.com; p. 13 Rickshu/Shutterstock.com; pp. 15, 24 aodaodaodaod/Shutterstock.com; p. 17 Naruedom Yaempongsa/Shutterstock.com; pp. 19, 24 Blend Images/Shutterstock.com; pp. 21 Paul Looyen/Shutterstock.com; pp. 23, 24 Antonina Potapenko/Shutterstock.com.

Printed in the United States of America

CPSIA compliance information: Batch #CW17GS: For further information contact Gareth Stevens, New York, New York at 1-800-542-2595.

Contents

Many sheep
live on farms.

Farmers take care
of them.

Sheep live
on lots of land.

They like to eat grass.

Dogs group
sheep together.
This is called herding.

Sheep grow lots of hair.
This is called wool.

Farmers cut wool
off sheep.

Wool is used
to make clothes!

Mother sheep
have babies.

Baby sheep
are called lambs.

Words to Know

clothes

lamb

wool

Index